Guide to the Trees, Shrubs, and Woody Vines of Tennessee

Guide to the Trees, Shrubs, and Woody Vines

of Tennessee

B. Eugene Wofford

Edward W. Chester

The University of Tennessee Press   Knoxville

Funds in support of this project were provided by the Center for Field Biology, a Tennessee Board of
Regents Center of Excellence at Austin Peay State University, Clarksville, Tennessee; the Hesler Fund,
Department of Botany, The University of Tennessee, Knoxville, Tennessee; and Discover Life in
America, Inc.

This book is printed on acid-free paper. This book was designed and typeset by Bill Adams on a Mac-
intosh computer system using QuarkXpress software. The text is set in Weiss 10/13, and display type
is set in Peignot Light. This book was manufactured by C&C Offset Printing Company, Ltd.

Frontispiece: Cumberland Rosemary (*Conradina verticillata* Jennison, Plate 103) is threatened federally
and in Tennessee, along with Appalachian Spiraea (*Spiraea virginiana* Britton, Plate 330). Photograph
by B. Eugene Wofford.

Library of Congress Cataloging-in-Publication Data

Wofford, B. Eugene.
Guide to the trees, shrubs, and woody vines of Tennessee / B. Eugene
Wofford and Edward W. Chester.—1st ed.
p. cm.
Includes bibliographical references (p.   ).
ISBN 1-57233-205-0 (pbk.: alk. paper)
1. Woody plants—Tennessee—Identification. I. Chester, Edward W. II. Title.
QK187 .W644 2002
582.16'09768—dc21   2002003607

To our families, our friends, field
biologists, and fellow
citizens of the state
of Tennessee

CONTENTS

Acknowledgments

THIS WORK would not have been possible without the cumulative efforts of past and present field botanists whose collections provide documentation for the known composition and distribution of the state's flora. To list them all would take too much space, and we would invariably omit some; however, we acknowledge with respect the dedicated work of each. We must especially acknowledge three botanical pioneers for their collections across the state and numerous other contributions to Tennessee floristics. Augustin Gattinger prepared the first Tennessee flora in 1901; Aaron J. Sharp reestablished the state herbarium (TENN) after it was destroyed by fire in 1934; Royal E. Shanks prepared the first checklist of Tennessee woody plants in 1952. Shanks and Sharp were mentors to dozens of botanists and their *Summer Key to Tennessee Trees*, published in 1950, is still in use.

Generic treatments, especially keys, were kindly supplied to us for three genera: *Crataegus* by Ron L. Lance (North Carolina Arboretum), *Ilex* by Ross C. Clark (Eastern Kentucky University), and *Viburnum* by Tim J. Weckman (Eastern Kentucky University). Richard J. Jensen, St. Mary's College, helped with the genus *Quercus*, especially the *Q. prinoides-muhlenbergii-michauxii* complex. In several cases, generic accounts were adapted and are cited at the appropriate generic treatment. Many other sources were generally used and are cited in the references section. We acknowledge and thank all of these botanists for their contributions, either directly or through their publications.

Alan Heilman, Department of Botany, UTK, shared his considerable expertise and spent many hours advising and helping with the photography. Joey Shaw, a graduate student in botany at UTK, provided yeoman's work in testing the keys, proofing the manuscript, and helping with photography. We acknowledge the following individuals/institutions for providing specimens and other contributions: David K. Smith, Dwayne Estes, Chris Fleming, Ron Lance, Janet Rock, V. E. McNeilus, Ken McFarland, James Donaldson, Ed and Meredith Clebsch, Julie Bland, Patricia Cox, Randy Small, Ed Lickey, Tim Weckman, Bill Wykle, R. Dale Thomas, Greg Schmidt, Jeff Walck, W. Michael Dennis, Zack Murrell, Tiffany Siler, Bryan Connolly, Paul Durr, Linda Moses, Keith Langdon, the Herbarium of the Great Smoky Mountains National Park, and the Botanical Research Institute of Texas.

Finally, we are deeply grateful for the thorough reviews and numerous helpful suggestions by Dr. R. Dale Thomas, professor of biology and curator of the herbarium, The University of Louisiana, Monroe, Louisiana, and Dr. Richard Jensen, professor of biology, St. Mary's College, and director of the Greene-Nieuwland Herbarium, University of Notre Dame, Indiana.

Introduction

TENNESSEE HAS been long recognized for its diversity in topography, geology, and especially its flora. Augustin Gattinger, generally considered to be the "Father of Tennessee Botany," commented upon his arrival in 1858 that Tennessee "possessed botanically and geologically so many and so diversified points of interest that a whole lifetime of a competent investigator could not exhaust and unravel all the problems and collect the various plants, minerals, and rocks" (Gattinger 1901). Contemporary botanists share these views (see discussions by more than 20 botanists in Chester (1989), and Wofford and Kral (1993)).

Factors accounting for the Tennessee floristic diversity have been variously elucidated by others (e.g., Griffith, Omernik, and Azevedo 1997; Luther 1977; Miller 1974; Sharp 1966, 1970), but will be mentioned here. The elongated east-west axis of the state crosses five major physiographic provinces and several provincial sections and subsections, each with a unique geological history. The mountainous eastern part is geologically old and has been available for occupancy and evolution of plants during times of great climatic shifts and biotic migrations. Non-mountainous western parts are much younger geologically and were covered by marine waters in the recent geological past. Between these two regions, central Tennessee includes a basin eroded from nearly pure limestone and surrounded by dissected uplands known for iron-bearing rocks westward and extensive karst topography east and west. Three major rivers and their meandering, complex floodplains and intricate and extensive drainage patterns affect each county. Thus, across the state there is much variability in topography, relief (about 200 to greater than 6000 feet), slope degree and aspect, substrates, and soils. Also, there are great climatic differences, expressed in variable temperatures, precipitation, wind currents, and length of growing season.

It is no surprise that Tennessee includes innumerable landscapes and microenvironments supporting a vascular flora exceeding 2800 taxa of diverse origins, ages, and relationships. As Graham (1965, 1999) and Sharp (1966, 1970) pointed out, this assemblage is large, diverse, and of tremendous interest for a number of reasons, including (1) the antiquity of many of its components, (2) the possible origin of many of its groups in Asia and Europe, (3) the presence of disjunct taxa with close relatives in Asia, Central America, and western, midwestern, and northern North America, (4) its interesting endemics, (5) the genetic complexity of many groups, and (6) its serving as a source of parent materials for areas to the south after receding marine waters and to the north as glaciers retreated.

Anthropogenic influences, mostly over the past 300 years, also must be mentioned. These include agriculture, the timber industry (including monoculture plantings), mining, stream channelization, high and low dams with resulting reservoirs, and urban sprawl. As a result, many natural communities/taxa have been decimated, but numerous weedy (and often non-native) taxa have flourished in the resulting disturbed habitats. Also, the introduction of diseases, damaging animals (especially insects), and exotic plant species are contributing factors in determining floristic composition.

Even today, the Tennessee flora is dynamic and not static. We are a "botanical border state" and regularly receive floristic elements from many migratory pathways and spill-over elements from adjacent provinces. In addition, field research is far from complete and our knowledge of the Tennessee flora is constantly growing. Each collecting trip away from the major points of study within the state (mainly the colleges and universities) brings back county records and often new reports for the state.

A Brief Historical Summary of Tennessee Botany

Tennessee vascular plants were first treated comprehensively by Gattinger (1901), who provided both an annotated list and a "philosophy of botany." Later treatments include the detailed account of ferns by Shaver (1954), a checklist of woody plants with distribution data by Shanks (1952a, supplemented 1953, 1954), and checklists of monocots and dicots by Sharp et al. (1956, 1960). Throughout and since that period, numerous papers, theses, and dissertations have significantly increased our knowledge of the state's flora (Bates 1985).

Recently, Wofford and Kral (1993) compiled an updated list of Tennessee vascular plants. County distribution maps for each taxon, based on the Wofford &

Kral list, were compiled by Chester et al. (1993) for pteridophytes, gymnosperms, and monocots; and Chester et al. (1997) for dicots. Current revisions, updated yearly, may be accessed at our website:

<www.bio.utk.edu/botany/herbarium/vascular/atlas.html>

or

<www.apsu.edu/biol_page/center.htm>

The first publication specifically designed for the identification of Tennessee's woody plants (in part) was that of Billings, Cain, and Drew (1937), who developed a winter key to trees of eastern Tennessee. Shanks and Sharp (1947) published a summer key to trees of the same area in the *Journal of the Tennessee Academy of Science.* This key was based at least in part on material originally prepared in 1937 by Stanley A. Cain as a classroom handout for introductory botany classes at The University of Tennessee, Knoxville. The key was expanded to include the entire state, copyrighted, and published by the Department of Botany, The University of Tennessee, Knoxville (Shanks and Sharp 1950) as the popular *Summer Key to Tennessee Trees.* Later, publication was assumed by The University of Tennessee Press (Shanks and Sharp 1963), and it now is in its ninth printing without revision.

Rationale and Objectives

Since 1947 the Shanks and Sharp dichotomous key has been the mainstay for teaching tree identification in our state. This publication has been used in innumerable classroom and field settings, and its influence on teaching, learning, and promoting botany in Tennessee has been immeasurable. However, only trees, small trees, and a few large shrubs (about 190 taxa) are included, and the numerous other woody plants must be sought in other sources. Also, distribution and descriptive data are not included, and since 1947 much additional information on the composition of the state's woody flora has accrued and numerous changes have resulted from taxonomic revisions and nomenclatural corrections. As currently known, slightly more than 400 (about 14.3 percent) of the state's vascular taxa are woody (trees, small trees, shrubs, subshrubs, vines). These taxa dominate most natural plant communities in the state, and their identification is the cornerstone of almost all botanical, ecological, and environmental studies. Thus, the need for an updated and expanded treatment, including all woody taxa and utilizing current taxonomy and nomenclature, is obvious. It is with those thoughts in mind that the present treatment was developed.

Our basic objective is to enable interested persons to identify the woody plants of the state. Certainly there is satisfaction (and often necessity) in identifying and knowing taxa, but that information can and should lead to additional interest in and knowledge about distributions, ecology, and relationships among taxa. With knowledge will come concern and appreciation, and only then can we sensibly conserve and preserve this vast and vital part of our natural heritage; that is our ultimate goal.

THE PHYSIOGRAPHIC REGIONS of TENNESSEE

Mississippi River Valley (MV)	Eastern Highland Rim (EHR)
Coastal Plain (CP)	Cumberland Plateau (CU)
Western Highland Rim (WHR)	Valley and Ridge (VR)
Central Basin (CB)	Unaka Mountains (U).

Five physiographic provinces occur in Tennessee (Fenneman 1938), including, from west to east: the Coastal Plain (consisting of the Mississippi River Valley and the Coastal Plain Upland), the Interior Low Plateau (Western Highland Rim, Central Basin, and Eastern Highland Rim), the Appalachian Plateau (Cumberland Mountains and Cumberland Plateau), the Valley and Ridge, and the Blue Ridge (Unaka Mountains). These provinces and sections are shown in map 1; the counties of Tennessee are shown in map 2. Our summary of each province is taken primarily from Griffith, Omernik, and Azevedo (1997), Luther (1977), and Miller (1974); designation of the three "Grand Divisions" of Tennessee (West, Middle, East) is from standard usage. Potential natural vegetation is from Küchler (1964). Flora and vegetation references include only the most recent papers. For other references, see Bates (1985) and papers cited by the numerous authors participating in the symposium on the flora and vegetation of Tennessee (Chester 1989). Also, Hackney, Adams, and Martin (1992) and Martin, Boyce, and Echternacht (1993a, 1993b) provide much additional information on biodiversity of the Southeast and Tennessee.

WEST TENNESSEE (W TN)

1. MISSISSIPPI RIVER VALLEY (MV). The Mississippi Valley, a Section of the Coastal Plain Province, is a broad alluvial plain up to 14 miles wide; river terraces and natural levees provide the major relief with average elevations about 250 feet above sea level. Boundaries are the Mississippi River on the west and Tertiary bluffs on the east. This is a relatively homogenous region of Quaternary alluvial deposits of sand, silt, clay, and gravel; regular flooding replenishes the sediments. Some clayey soils with poor drainage contain swamps and oxbow lakes, i.e., Reelfoot Lake (Lake and Obion Counties) formed by an earthquake in the early 1800s.

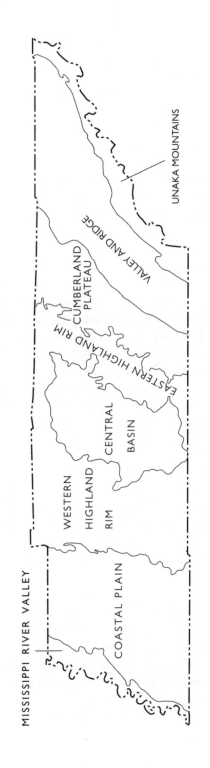

Map 1. A Generalized Physiographic Map of Tennessee

Potential natural vegetation is southern floodplain forest of oak-tupelo-bald cypress. Braun (1950) included the region within the Mississippi Alluvial Plain of the Southeastern Evergreen Forest Region, an area of swamps and bottomland forests. Sharitz and Mitsch (1993) discuss various aspects of these forests. Today, better-drained areas are almost totally in tilth and much of the presettlement flora is undocumented. Guthrie (1989), Heineke (1989), Henson (1990), and Miller and Neiswender (1989) provide information on the current plant life.

2. COASTAL PLAIN (CP). The West Gulf Coastal Plain Section of the Coastal Plain Province is sometimes referred to as the Plateau Slope in Tennessee. It extends from the MV to (or nearly to) the Tennessee River and is heterogenous in geology and topography; three distinct areas are recognized. Steep bluffs bordering the MV are referred to as Bluff Hills; the most prominent are known as the Chickasaw Bluffs. These bluffs extend about 100 feet above the MV and are capped by loess that may exceed 60 feet in depth. The nearly flat West Tennessee Plain (Loess Plains of Griffith, Omernik, and Azevedo 1997) extends north-south across the state from the bluffs eastward to the headwaters of streams that drain westward toward the Mississippi River and those that drain eastward toward the Tennessee River. This region of thick loess on gently rolling plains (250–500 feet in elevation) is broken only by several large rivers and almost all is in tilth. Eastward the Coastal Plain Uplands, or Southeastern Plains and Hills (roughly the Tennessee River drainage), consists of uplands capped with Cretaceous gravels and sands. Elevations range from 400 to >700 (average about 500) feet and the topography is more rolling than that of the plains to the west.

Potential natural vegetation is oak-hickory forest on uplands (most of the area), southern floodplain forest along tributaries of the Mississippi River (Sharitz and Mitsch 1993), some oak-hickory-pine forest to the south, and small sections of prairie-oak-hickory forest mosaic to the north. Braun (1950) placed the Tennessee CP within the Mississippi Embayment Section of the Western Mesophytic Forest Region, while Bryant, McComb, and Fralish (1993) include it under Oak-Hickory Forests, which includes Braun's Western Mesophytic and Oak-Hickory Forest Region. The diverse area now includes a mosaic of unlike vegetation types, such as oak-hickory and swamp forests, mixed mesophytic communities, and prairie-barren remnants [DeSelm and Murdock (1993) provide an overview of grass-dominated communities within the state]. Almost all of West Tennessee is productive agriculturally, but most natural plant communities have been depleted except for some bottomlands. Heineke (1989), Lewis and Browne (1991), and Miller and Neiswender (1989) present data on plant communities and flora of the area.

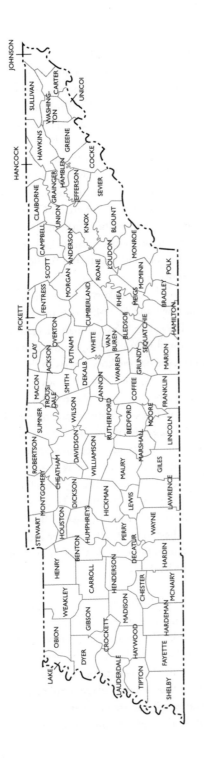

Map 2. The 95 Counties of Tennessee

Middle Tennessee (M TN)

3. WESTERN HIGHLAND RIM (WHR). This subsection of the Highland Rim Section, Interior Low Plateaus Province, is bounded on the east by the Central Basin Subsection and extends westward to the Coastal Plain Province at or just west of the Tennessee River. The WHR is one of the largest of Tennessee's regions and is basically a broad plateau tilted from east to west with a strongly dissected, rolling terrain and numerous stream valleys; elevations are 500 to 1000 feet. The area has developed primarily on St. Louis and Warsaw limestones (Mississippian); cherty Cretaceous gravel caps some ridges. Northeastern sections of Stewart County, northern sections of Montgomery and Robertson Counties, and northwestern sections of Sumner County are within the southern Pennyroyal ("Pennyrile") Plain Subsection, a nearly flat region known for its karst terrain and lying mostly in Kentucky; this subsection is included with the WHR in our discussion of distributions and floristic relationships.

Potential natural vegetation is mostly oak-hickory forest with some oak-hickory-pine forest to the south and a narrow band of prairie-oak-hickory forest (the Big Barrens Region) to the north. This is the Mississippian Plateau Section of the Western Mesophytic Forest Region of Braun (1950), a mosaic of unlike communities basically transitional between the oak-hickory forests to the west and mixed mesophytic forests to the east; however, oak-hickory types prevail, and mixed mesophytic conditions are circumscribed and rare. Bryant, McComb, and Fralish (1993) place this area within the Oak-Hickory Forests (combining Western Mesophytic and Oak-Hickory Forest Regions of Braun). This is the iron-mining region of Tennessee, and most of the forests were degraded from the early 1800s to early 1900s in the mining and smelting process and as a result of general lumbering and agriculture; extensive secondary forests cover much of the area now (Wayne County is the most heavily forested county in Tennessee). Accounts of present-day vegetation and flora may be found in Baskin, Baskin, and Chester (1994, 1999), Baskin, Chester, and Baskin (1997), Carpenter and Chester (1987), Chester (1993), Chester, Jensen, and Schibig 1995, Chester et al. (1997), Joyner and Chester (1994), and Souza and Kral (1990).

4. CENTRAL BASIN (CB). This section of the Interior Low Plateaus Province is often divided into Outer and Inner Basin components. The outer encircling knobby region includes strong intrusions from the surrounding Highland Rim. Elevations are from 500 to 1200 feet (average 750 feet). The inner basin is less hilly and lower in elevation (average 600 feet) with numerous outcrops of Ordovician limestones, often in flat sheets and constituting typical cedar glade topography.

The potential natural vegetation of the inner basin is that of cedar glades (*Juniperus-Quercus-Sporobolus*), while the outer basin is oak-hickory forest. Braun

(1950) referred to the CB as the Nashville Basin Section of the Western Mesophytic Forest Region and noted that cedar glades are most distinctive. Baskin and Baskin (1989, 1999) and Quarterman, Burbanck, and Shure (1993) provide overviews of cedar glades, including physiographic, vegetational, and floristic features. Other recent accounts of the basin vegetation-flora include Crites and Clebsch (1986), McKinney and Hemmerly (1984), Schibig (1996), and Somers (1986).

5. EASTERN HIGHLAND RIM (EHR). The EHR, a subsection of the Highland Rim Section, Interior Low Plateaus Province, is a continuation of the WHR, an upland area surrounding the CB. We basically follow Griffith, Omernik, and Azevedo (1997) in determining northern and southern boundaries. To the north, the EHR ends where the Pennyroyal Plain begins, except we include all of Sumner County within the WHR; the EHR begins with Macon County. Southward, the narrow extension of the Outer Basin is the boundary; Giles County is WHR, Lincoln is EHR.

The EHR is much narrower than the WHR, averaging about 25 miles in width. Topography is quite intermediate between the Cumberland Plateau to the east and the CB. There is more level (tableland) terrain than is found on the WHR, but also more deeply dissected and rugged terrain. A nearly flat area in the south (Coffee, Cannon, and Warren Counties) is known as the Barrens. Elevations mostly range from 800 to 1300 feet (average 1000); Short Mountain (Cannon County) is over 2000 feet. Most of the area is underlain by Mississippian limestones; an extensive karst region occurs in Warren and White Counties, and caves and sinkholes occur throughout.

The potential natural vegetation is oak-hickory forest. Braun (1950) placed the area within the eastern part of the Western Mesophytic Forest Region, while it is part of the Western Mesophytic/Oak Hickory Forests combination of Bryant, McComb, and Fralish (1993). Some mixed mesophytic communities occur, as do upland swamps and bottomland and hemlock forests (McKinney 1986, 1989). The southern EHR is well known for several natural grasslands, swamps, and wet woodlands harboring many rare plants, several with Coastal Plain affinities.

EAST TENNESSEE (E TN)

6. CUMBERLAND PLATEAU (CU). The CU is within the Appalachian Plateau Province and consists of low mountains northward (the Cumberlands) and tablelands averaging 1000 feet higher than the EHR to the west or the Great Valley to the east. Elevations are up to 3500 feet, averaging 1700–1900 feet. Much of the plateau is rimmed by cliffs broken by narrow, steep-walled gorges, stream valleys, and ravines that run back into the tablelands. Sandstone cliffs and waterfalls, such as Falls Creek Falls, are characteristic features; northward, sandstone rockhouses and natural bridges are unique. Considerable areas of the tablelands are poorly drained and

include a number of wetland types such as floodplains, swamp forests, natural ponds, and wet meadows. The Sequatchie Valley is a major feature from about mid-state southward, while the smaller Elk Valley is northward. The geology is Pennsylvanian-age conglomerate, sandstone, siltstone, and shale. Coal beds occur throughout the Plateau but are more prevalent northward and are a valuable resource, although the results of mining often have been environmentally devastating.

The potential natural vegetation is mixed mesophytic forest (*Acer-Aesculus-Fagus-Liriodendron-Quercus-Tilia*), oak-hickory-pine, and oak-hickory forest. Rock outcrop communities are not uncommon (Quarterman, Burbanck, and Shure1993). Braun (1950) included the area within her Mixed Mesophytic Forest Region, which reached best development in the Cumberland Mountains. Logging, strip mining, agriculture, fire, chestnut blight, and other anthropogenic factors have altered most of the forests. Hinkle et al. (1993) provide an overview of the area. Studies of current plant life can be found in Caplenor (1955, 1965), Clements and Wofford (1991), Hinkle (1989), Jones (1989), Quarterman, Turner, and Hemmerly (1972), Schmalzer (1989), Shaw (2001), and Wofford et al. (1979).

7. VALLEY AND RIDGE (VR). This region, also called the Great Valley of East Tennessee, is a relatively low-lying area between the CU to the west and the Unaka Mountains to the east. The general landscape is that of a broad rolling upland with numerous ridges, ravines, and valleys. Elevations range from 700 to just over 3000 feet. The bedrocks include Silurian, Ordovician, and Cambrian-aged limestone, shale, siltstone, sandstone, chert, and marble.

Braun (1950) placed the VR in the Ridge and Valley Section of the Oak-Chestnut Forest Region, while more recent treatments (Stephenson, Ash, and Stauffer 1993) include it within Appalachian Oak Forests. The potential natural vegetation is Appalachian oak forests; American chestnut is rarely found today. Martin (1989) provides an overview of the existing plant communities.

8. UNAKA MOUNTAINS (U). Although geographers and geologists are not uniform in their interpretation of mountain names, many call that portion of the Appalachian Mountains extending from Maryland to Georgia the Blue Ridge. Early geologists referred to the Tennessee mountains separating the Great Valley from North Carolina as the Unaka Range, and we follow that terminology here. Most individual mountains within the range have names, as do groups of mountains; best known are the Great Smoky Mountains and the national park that encompasses them. Others include Chilhowee, English, and Roan. Elevations generally range from1000 to >6000 feet with Clingmans Dome, the highest point in Tennessee, reaching 6643 feet. These mountains are extremely old geologically with a mixture of bedrocks, including igneous, metamorphic, and sedimentary

types. The region is characterized by steep, often forested slopes, cool, clear streams, and rugged terrain. Some of the best-known landscapes are the relatively flat, mountain encircled coves (e.g., Cades, Tuckaleechee, Wear).

The potential natural vegetation is Appalachian oak forest, northern hardwoods (*Acer-Betula-Fagus-Tsuga*), and southeastern spruce-fir forest (*Picea-Abies*) at highest elevations. For Braun (1950), the area included the Southern Appalachian Section of the Oak-Chestnut Forest Region with mixed mesophytic or cove hardwoods (Hinkle et al. 1993), oak-pine, northern hardwoods, and spruce-fir elements. In addition, there are shrub, grass, and heath balds. Stephenson, Ash, and Stauffer (1993) and White et al. (1993) provide overviews of this diverse region; Clebsch (1989), Murrell and Wofford (1987), Ramseur (1989), Thomas (1989), Warden (1989), White (1982), and Wofford (1989a, 1989b) provide insight into the flora and vegetation.

PHysiogRApHic-FlorisTic RElATioNsHips

As expected, there is a strong relationship between physiographic regions and the occurrence of taxa. Shanks (1952b, 1958) earlier pointed this out, although his designation of floristic regions often involved broad categories, e.g., "bottomland" and "mountain" flora, as well as U.S. Forest Service types. An examination of currently known distributions of Tennessee woody plants (Chester et al. 1993, 1997, Austin Peay State University Center for Field Biology and University of Tennessee Herbarium website) shows distinct correlations between distribution and physiography, and we limit the discussion here to those patterns.

The distribution of Tennessee's woody flora within the physiographic provinces and their major subdivisions is shown in Appendix 1 (for species and lesser taxa) and in Appendix 2 (for genera). These tables are based on voucher specimens used in the *Atlas of Tennessee Vascular Plants* (Chester et al. 1993, 1997, Austin Peay State University Center for Field Biology and University of Tennessee Herbarium website) and may be modified in the future as additional fieldwork provides new information. Also, it should be noted that there is certain margin of error when physiographic province is designated from the often incomplete data on herbarium specimens. This is especially true (but not limited to) the transition area between the MV and the CP, and between the EHR and the CU.

No taxa are limited to the MV (*Gleditsia aquatica, Schisandra glabra,* and *Ulmus crassifolia* are nearly limited there), but several taxa are essentially confined to the MV and CP, occasionally extending east of the Tennessee River. These include *Brunnichia ovata, Carya aquatica, C. illinoinensis, Fraxinus profunda, Gleditsia aquatica, Nyssa aquatica, Planera aquatica, Quercus texana, Styrax americanus, Taxodium distichum, Trachelospermum difforme, Vitis cinerea* var. *cinerea,* and *V. palmata.*

Only *Ribes odoratum* (totally) and *Crataegus mollis* (mostly) are confined to the WHR. However, there is a strong spillover from the CP, especially in bottom-lands near the Tennessee River (Webb and Bates 1989, also see above list). Also, there is a strong CB element, especially along the Cumberland River, which apparently has served as a migratory pathway for such species as *Clematis cates-byana, Forestiera ligustrina, Fraxinus quadrangulata, Philadelphicus pubescens* var. *intectus,* and *Ulmus thomasii*. In addition, there is a small Appalachian element along the Ten-nessee River, i.e., *Halesia tetraptera* and *Rhododendron maximum*. Upland wet woods on the Pennyroyal Plain support a few species rarely found elsewhere in Middle Ten-nessee, most noticeably *Leucothoe racemosa* and *Populus heterophylla*.

While the CB and its cedar glades harbor numerous endemic or near endemic herbs (Baskin and Baskin 1989, 1999), few woody plants are restricted to the CB. *Vitis rupestris* has not been seen since Gattinger's pre-1900 collection and is now considered to be possibly extirpated. Likewise, *Lonicera reticulata* is known only from a pre-1900 collection. One of the rarest woody plants in the United States occurs in the CB; *Crataegus harbisonii* is known in Tennessee from only one recently rediscovered specimen growing in its type locality in Davidson County (Lance and Phipps 2000). Several taxa essentially confined to the CB (most intrude onto the surrounding WHR and/or EHR) include *Clematis catesbyana, Forestiera ligustrina, Lonicera reticulata, Neviusia alabamensis, Philadelphicus pubescens* var. *intectus, Rhamnus lance-olata* (with an outlier in Claiborne County), *Rosa virginiana,* and *Ulmus thomasii*.

Woody plants limited (or mostly limited) to the EHR include *Clethra alnifolia, Croton alabamensis, Gaylussacia dumosa, Prunus pumila, Vaccinium elliottii,* and *Viburnum molle*. Numerous taxa most often found in the CB extend onto the EHR (see list under CB). Also, numerous taxa more often found on the CU (or eastward) occur on the EHR, including *Aristolochia macrophylla, Cornus alternifolia, Lonicera dioica, Mag-nolia tripetala, Rhododendron maximum,* and *Tsuga canadensis*.

Taxa essentially limited to the CU include *Buxella brachycera, Comptonia peregrina, Conradina verticillata, Cotinus obovatus, Polygonella americana, Ribes curvatum, Taxus canadensis,* and *Viburnum bracteatum*. A considerable number of species constitutes the "mountain flora" (Shanks 1958), or taxa mostly occurring from the CU east-ward. Examples are *Acer pensylvanicum, Betula alleghaniensis, B. lenta, Cornus alternifolia, Corylus cornuta, Pinus rigida, P. strobus, Magnolia tripetala, Rhododendron maximum, Thuja occidentalis,* and *Tsuga canadensis*. Only *Paxistima canbyi* and *Rhamnus alnifolia* are lim-ited to the VR, but numerous elements of the "mountain flora" occur there.

The Unakas, with their several unique plant community types (Clebsch 1989) and diverse residual elements of the once widespread Arcto-Tertiary Geoflora (Wofford 1989a), probably are the best-known of Tennessee's floristic regions. Numerous Tennessee taxa are found there only; several are southern Appalachian endemics (Wofford 1989a), indicated by an asterisk in the following list: *Abies* fraseri*, Alnus viridis ssp. crispa, Betula cordifolia, Buckleya distichophylla*, Diervilla rivularis*, Gaylussacia ursina*, Hydrangea radiata*, Kalmia angustifolia var. carolina, Leiophyllum buxifolium, Leucothoe recurva*, Linnaea borealis, Lonicera canadensis, Menziesia pilosa, Nemopanthus collinus*, Picea rubens, Pieris floribunda, Potentilla tridentata, Prunus pensylvanica, Prunus virginiana, Rhododendron viscosum, Ribes glandulosum, R. rotundifolia, Rubus idaeus ssp. strigosus, Sorbus americana, Spiraea alba, Tsuga caroliniana*, Vaccinium hirsutum*, and Viburnum lantanoides.

COMPARATIVE RICHNESS ACROSS THE STATE

The Grand Divisions are quite unequal in number of native taxa at both the genus and species/lesser taxa levels (table 1), with richness increasing from west to east. Percentage wise, about 69 percent of native genera known from the state occur in W TN, 85 percent in M TN, and 94 percent in E TN. At the species/lesser taxa level, 56 percent of the native taxa occur in W TN, 80 percent in M TN, and 93 percent in E TN.

Distribution of taxa within the physiographic provinces/divisions is shown in table 2. As expected, the MV, the smallest unit in area and probably the most disturbed, is the least floristically rich. Interestingly, the CB, one of the botanically better-known areas of the state, is comparatively low in woody taxa richness. The rims (WHR, EHR) and the VR are close in richness, even though the EHR is the smallest of the three. As further suspected, the eastern mountains (CU and U) are the richest; perhaps surprisingly, the CU exceeds the U by one genus and 10 species.

GEOGRAPHIC AFFINITIES OF THE WOODY FLORA

The woody plants are especially well suited for showing geographic affinities because range maps are available for many species, notably trees. Various map sets were used to provide insight into the floristic affinities of the state's flora (Little 1971, 1977, 1981; Flora of North America 1993, 1997). Also, ranges given by Gleason and Cronquist (1991) and other references cited with specific genera were used to obtain the following specific examples.

Much of the flora is intraneous, or well within the total range of given taxa (Amelanchier arborea, Juniperus virginiana, Sambucus canadensis). However, several other geographic elements contribute significantly to the state's flora.

The distribution of a large number of taxa lies primarily to the north of Tennessee. Some of this northern element are widely distributed within the state (Acer saccharum ssp. saccharum, A. saccharum ssp. nigrum, Carya laciniosa, Celtis occidentalis, Euonymus atropurpureus, Fraxinus quadrangulata, Juglans cinerea, Pinus virginiana, Populus

Table 1

Number of Native Genera and Species/Lesser Taxa within the Three Grand Divisions of Tennessee

Group	West Tennessee (MV, CP)	Middle Tennessee (WHR, CB, EHR)	East Tennessee (CU, VR, U)
Genera (143)	99	121	135
Species/Lesser Taxa (358)	203	279	331

Table 2

Number of Native Genera and Species/Lesser Taxa within the Physiographic Provinces of Tennessee

Group	MV	CP	WHR	CB	EHR	CU	VR	U
Genera (143)	81	98	109	97	108	119	112	118
Species/Lesser Taxa (358)	152	200	244	200	238	271	251	261

Table 3

Taxonomic Summary for the Tennessee Woody Flora

Group	No. Families	No. Genera	No. Intro. Species/ Lesser Taxa	No. Native Species/ Lesser Taxa	Total No. Species/ Lesser Taxa
Gymnosperms	4	8	—	14	14
Angiosperms:					
Monocots	3	3	—	7	7
Dicots	63	150	45	337	382
Totals	70	161	45	358	403

grandidentata, Quercus bicolor, Q. macrocarpa, Q. palustris); others show an Appalachian extension representing a boreal invasion during glaciation (see previous list, also Wofford 1989a). A few northern taxa extend onto the Cumberland Plateau (*Comptonia peregrina*), while others, constituting a "mountain element," [see previous list and/or Shanks (1958)] are found in both the Unakas and on the Cumberland Plateau (occasionally in the Valley and Ridge as well).

Some species are found mostly to the east of Tennessee and reach their western limits within the state (e.g., *Kalmia latifolia, Oxydendrum arboreum*), while others are distributed primarily west-northwest of Tennessee (e.g., *Cornus drummondii, Gymnocladus dioica, Prunus mexicana, Salix exigua*).

A southern element is well-represented, especially in western sections. Taxa whose distributions lie mostly within the Mississippi Embayment and eastward onto the Atlantic Coastal Plain, but extend northward into West Tennessee (mostly) include *Carya aquatica, Forestiera acuminata, Nyssa aquatica, Planera aquatica, Styrax grandifolius*, and *Taxodium distichum*. Some taxa found mostly in West Tennessee are primarily distributed in the Mississippi Embayment and westward (e.g., *Carya illinoinensis, Gleditsia aquatica, Quercus texana, Ulmus crassifolia*). Some southern species are found across the state, mostly in southern counties (e.g., *Callicarpa americana, Hydrangea quercifolia, Magnolia virginiana*), while others are rather widely distributed within the state (*Carya pallida, Quercus nigra, Q. pagoda*). Also, there is a southern, mostly Coastal Plain element, that includes some of our most rare species (*Clethra alnifolia, Croton alabamensis, Nestronia umbellula, Vaccinium elliottii*).

Floristic Diversity

Taxonomic Summary

The Tennessee woody flora consists of 403 species and lesser taxa (358 native, 45 naturalized) within 70 families and 160 genera. This amounts to about 14.3 percent of the state's flora (15.8 percent of the native taxa and 8.1 percent of the introduced element). A summary of the distribution of taxa within major groups is given in table 3.

Large families, based on the number of species and lesser taxa, are Rosaceae (63), Ericaceae (36), Fagaceae (24), Caprifoliaceae (24), Fabaceae and Saxifragaceae (including Grossulariaceae and Hydrangeaceae) (15 each), Salicaceae (14), Vitaceae (13), Juglandaceae (11), and the Aceraceae, Betulaceae, Oleaceae, Pinaceae, and Ulmaceae (10 each). Two families have 8 taxa, 3 have 7, 2 have 5, 5 have 4, 7 have 3, 13 have 2, and 25 have 1.

Large genera, based on the number of species and lesser taxa, are *Quercus* (21), *Crataegus* (14), *Prunus* and *Rubus* (12 each), *Acer, Rhododendron, Viburnum,* and *Vitis* (10 each), *Carya, Salix,* and *Vaccinium* (9 each), *Hypericum* (8), *Lonicera* (7), *Ilex, Magnolia, Pinus, Rosa,* and *Ulmus* (6 each), and *Cornus, Populus, Ribes,* and *Spiraea* (5 each). In addition, 9 genera have 4 taxa, 10 have 3, 26 have 2, and 94 have 1. The large number of genera with few taxa (58 percent of genera include only one taxon, nearly 81 percent include fewer than four taxa) is an indicator of diversity; Balmford, Jayasuriya, & Green (1996) point to increasing awareness of the correlation between generic richness and species diversity of an area.

Listed Elements

Based on the rare plants list of the Tennessee Natural Heritage Program (2001), 55 woody taxa are elements of concern federally and/or in the state (table 4). This is about 14 percent of woody taxa; overall, about 16 percent of all Tennessee vascular taxa are listed. Nineteen taxa are Special Concern (S), 19 are Threatened (T), and 17 are Endangered (E). Two taxa listed as Threatened by the state also are federally listed as Threatened (LT). Also, five of the taxa given in table 4 are possibly extirpated (PE) in Tennessee:

(1) *Croton alabamensis* E. A. Smith *ex* Chapman, Alabama Croton, is a globally rare shrub apparently extant in two river valleys in Tuscaloosa and Bibb Counties, Alabama (Farmer and Thomas 1969, Clark 1971). The Tennessee record is based on a specimen at the University of North Carolina Herbarium with the following data (specimen not seen by us, cited by Farmer and Thomas 1969): "Tullahoma, Coffee County, Tennessee, T. G. Harbison 725, Aug. 10, 1899." Ross Clark, Eastern Kentucky University (personal communication, 1999) also has examined this specimen. There have been no other reports from Tennessee. The var. *texensis* Ginzbarg, endemic to Texas, was only recently described (Ginzbarg 1992).

(2) *Kalmia angustifolia* L. var. *carolina* (Small) Fernald, Sheep Laurel/Lambkill, is a globally rare taxon known only from Virginia, North and South Carolina, Georgia, and Tennessee (Kartesz and Meacham 1999). We have collections from Shady Valley Bog in Johnson County, but none since 1951.

(3) *Linnaea borealis* L., Twinflower, is an evergreen, trailing vine circumpolar in distribution with extant populations occurring southward as far as West Virginia. The Tennessee record was collected on 13 August 1892, from "Sevier County-in mountain woods" by the Knoxville educator and botanist Albert Ruth. The specimen was originally misidentified and remained in the private collection of Ruth until after his death. Then, a number of specimens were obtained from the estate by Dr. A. J. Sharp, who correctly identified the plant and first noted its importance. The specimen today is at TENN. White (1981) recounts the history of this collection, notes that there is some doubt that it was actually

collected in Tennessee, and details extensive efforts over the years to locate extant populations in East Tennessee.

(4) *Lonicera reticulata* Raf. [*L. prolifera* (Kirchn.) Rehd.], Grape Honeysuckle, is a wide-ranging species extending from Nova Scotia southward to Tennessee (Kartesz and Meacham 1999). Previous reports from the state include that of Small (1933), without specific locality, and Shanks (1952a) and Sharp et al. (1960) from Roane County. Vouchers for the latter reports have been determined to be *L. dioica*. Duncan (1967) mapped the species from Davidson County, apparently based on a specimen in the Gray Herbarium collected pre-1900 in "Davidson Co., Nashville" (data from the Tennessee Natural Heritage Program). We have not seen the specimen and our inclusion of this taxon is based on the Duncan report.

(5) *Vitis rupestris* Scheele, Sand/Sugar Grape, is known only from pre-1900 collections by A. Gattinger from Davidson County. Further examination of the specimens vouchering our previous reports from Hickman, Maury, and Wilson Counties (Chester, Wofford, and Kral 1997) indicates that these specimens represent other species of *Vitis*. Gattinger's voucher specimens at TENN and GH were annotated by M. O. Moore, who notes that the species was distributed once from south-central Texas to southwestern Pennsylvania but apparently has been extirpated from most regions; extant populations may occur only in Arkansas and Missouri (Moore 1991).

Introduced Taxa

A number of exotic species contribute to the state's woody flora. Several, mostly introduced from Europe and Asia as ornamentals, for wildlife food, for erosion control, or for other and often obscure reasons, have become serious pests, often replacing native vegetation; examples include *Albizia julibrissin, Ailanthus altissima, Lespedeza bicolor, Ligustrum sinense, L. vulgare, Lonicera japonica, L. maackii, Paulownia tomentosa, Pueraria montana* var. *lobata, Rosa multiflora,* and *Spiraea japonica*. Yet other exotic taxa persist and/or sparingly spread from plantings around old home sites and in cemeteries; examples are *Hedera helix, Malus pumila, Populus alba, Salix alba, S. babylonica,* and *Vinca minor*. Also, a few species native to other parts of North America have been introduced and are naturalized or are long persisting after cultivation, e.g., *Catalpa bignonioides, Maclura pomifera,* and *Magnolia grandiflora*.

Several species native to one part of the state are widely cultivated in other parts and often spread after planting, thus making statements about their natural distribution difficult. Among others, these include *Carya illinoinensis* (cultivated for fruit), *Catalpa speciosa* (as an ornamental and a forage plant for larvae used for fish bait), *Hydrangea quercifolia, Robinia hispida,* and *Lonicera sempervirens* (ornamentals), *Pinus taeda* (for erosion control and pulpwood), and *Robinia pseudoacacia* (for erosion control and fence posts).

Table 4

STATE AND/OR FEDERAL LISTED WOODY TAXA IN TENNESSEE

Species	Status
Abies fraseri	T[b]
Acer saccharum ssp. *leucoderme*	S[c]
Alnus viridis ssp. *crispa*	S
Amelanchier sanguinea	T
Berberis canadensis	S
Betula cordifolia	E[a]
Buckleya distichophylla	T
Castanea dentata	S
Clethra alnifolia	T
Comptonia peregrina	E
Conradina verticillata	T, LT[e]
Cotinus obovatus	S
Crataegus harbisonii	E
Croton alabamensis	E–PE[d]
Diervilla lonicera	T
Diervilla rivularis	T
Euonymus obovatus	S
Fothergilla major	T
Gaylussacia dumosa	T
Gelsemium sempervirens	S
Juglans cinerea	T
Kalmia angustifolia var. *carolina*	E–PE
Leucothoe racemosa	T
Linnaea borealis	E–PE
Lonicera canadensis	S
Lonicera dioica	S
Lonicera flava	S
Lonicera reticulata	E–PE
Magnolia virginiana	T
Menziesia pilosa	S
Nemopanthus collinus	S
Nestronia umbellula	E
Neviusia alabamensis	T
Paxistima canbyi	E
Pieris floribunda	T
Polygonella americana	E
Populus grandidentata	S
Potentilla tridentata	S
Prunus pumila	T
Prunus virginiana	S

Table 4 *(continued)*

Species	Status
Rhamnus alnifolia	E
Ribes odoratum	T
Schisandra glabra	T
Spiraea alba	E
Spiraea virginiana	E, LT
Symplocos tinctoria	S
Taxus canadensis	E
Thuja occidentalis	S
Tsuga caroliniana	T
Ulmus crassifolia	S
Vaccinium elliottii	E
Vaccinium macrocarpon	T
Viburnum bracteatum	E
Vitis rupestris	E–PE
Zanthoxylum americanum	S

NOTES: [a]Endangered, [b]Threatened, [c]Special Concern, [d]Possibly Extirpated, [e]Listed Threatened (unlike the other status listings, which are designated by the state of Tennessee, this is a federally designated status listing).

TAXONOMIC TREATMENT

The taxonomic treatment follows Wofford and Kral (1993) and the Tennessee Atlas (Chester et al. 1993, 1997). The following, more recent, treatments were adapted: *Flora of North America* [FNA] (1993) for gymnosperms; FNA (1997) for the Magnoliophyta: Magnoliidae and Hamamelidae; Isely (1990, 1998) for Fabaceae; and Luteyn et al. (1996) for Ericaceae. In cases where a specific generic revision or state treatment was used, that reference is cited with the genus treatment. Various other, generally standard sources were consulted and were invaluable as we attempted to interpret and summarize this material on the woody flora of Tennessee. These include Clark (1971), Duncan (1967), Duncan and Duncan (1988), Fernald (1950), Foote and Jones (1989), Gleason and Cronquist (1991), Godfrey (1988), Harlow et al. (1996), Kartesz (1994), Kartesz and Meacham (1999), Radford, Ahles, and Bell (1968), Smith (1994), Swanson (1994), and Wofford (1989b). Robertson (1974) was consulted extensively for genera of Rosaceae. Author names or their abbreviations follow Brummitt and Powell (1992). In all cases our ultimate source of information was the plants themselves, especially the collections at TENN.

FORMAT

This treatment is divided into three groups: (1) Gymnosperms, with the key to genera followed by the respective and alphabetical generic treatments and keys to species where needed, (2) Angiosperms-monocots, and (3) Angiosperms-dicots (generic keys divided into seven sections). Monocot and dicot genera are then given alphabetically, with keys to species.

For each taxon (species and lesser taxa) we provide (1) the scientific name and authorship (introduced taxa are indicated with an asterisk); (2) at least one generally accepted (used regularly in Tennessee) common name; (3) a brief description of flowers, the inflorescence type, and fruit type, including season of occurrence; (4) general habitat preferenda; (5) frequency of occurrence based on county distributions and ranging from common (documented from nearly all 95 counties), frequent (known from >50 percent of counties), occasional (known from <50 percent of counties), infrequent (generally rare or known from fewer than 10 counties); (6) distribution in the state as statewide (known from some part of each of the five physiographic provinces), by Grand Division (known from all provinces and/or recognized sections/subsections within that Division), by physiographic province, or by a combination of these; (7) listed status; (8) origin of introduced taxa; (9) occasional comments; and (10) pertinent synonymy. Cultivated taxa that are not naturalized within the state are not included in the keys. A list of these taxa represented by collections is given in appendix 3.

Our keys are based on vegetative material wherever possible. In several cases, determination requires flowering/fruiting specimens and while that requirement exacerbates the often laborious and tedious task of using a dichotomous key, and frustrates all of us as well, it is simply a fact of botanical life that some taxa cannot be separated with vegetative characters alone. However, we have attempted to provide a treatment that can be used at the beginning level and that will be a valuable resource for the advanced student as well. The keys should be used in conjunction with the distribution data available in the *Atlas* (see Austin Peay State University Center for Field Biology and University of Tennessee Herbarium website). A 10x hand lens is required in many cases.

Do not expect perfection in these keys. Rarely will a simple key encompass the morphological variability shown within and between most taxa. Also, our interpretations of variability, and the translation of those interpretations to printed keys and treatments are regularly, but not embarrassingly, inadequate; our recognized inadequacies are incentive for more work in the field, herbarium, and laboratory. Likewise, changes in composition and distribution of the state's flora, as well as in taxonomy and nomenclature, are inevitable. Like probable shortcomings in our work, these can be addressed by ourselves and/or other field botanists in future editions.

Photography

Unlike most other woody plant treatments, we chose to illustrate specimens in a one-plane format. We believe this presentation has the advantage over habit shots by enabling the user to view the entire leaf and easily determine its arrangement, type, margin characteristics, relative length, width, etc. Unfortunately, photographs of this type often make it difficult to determine if the leaves are arranged in one or more than one plane. Also, we have not included a size scale and size usually cannot be determined from the photographs (specific leaf sizes often are given in the keys where this is a diagnostic feature). Overall, we believe the advantages of this method outweigh the disadvantages, and we hope we have succeeded in our attempt. Even though our keys are based almost solely on vegetative characters, we occasionally included reproductive structures as a further aid in identification. Also, for a few species we have included insets of surface features critical to identification.

Photographs were made, when possible, from freshly collected and pressed specimens. The remainder were from herbarium specimens and we regret the occasional inclusion of mounting media, damaged/faded leaves, and other distractions. In a few cases, we used out-of-state collections when our material was either unavailable or unsuitable for use. We omitted 13 taxa that were either unavailable, unsuitable, or if their identity by macro-photography could not be elucidated from closely related taxa.

GLOSSARY

Abaxial: away from the axis; when referring to a leaf, the lower surface.

Achene: small, dry, 1-locular, 1-seeded indehiscent fruit with the seed coat and ovary wall separate. Plate 97.

Acicular: needle-shaped.

Acorn: the 1-seeded fruit of oaks; consisting of a cup-like base and the nut. Plates 261–270.

Actinomorphic: having a symmetry such that two or more median longitudinal divisions through the flower will yield identical paired mirrored images; sometimes referred to as radial symmetry.

Acuminate: a tip whose sides are variously concave and tapering to a point. Plates 67, 87, 141.

Acute: sharply ending in a point with margins straight or slightly convex. Plate 349.

Adnate: the fusion or growing together of unlike parts.

Aerial: in the air, as in vines that attach to other objects with above ground parts.

Aggregate fruit: a compound fruit derived from the coherence of 2 to many simple, superior ovaries of a single flower; e.g., blackberry (Rosaceae).

Alternate: borne singly and not opposite, i.e., one leaf at a node. Plates 55, 345.

Anastomosing: rejoining to form a network.

Apex: the tip of a structure.

Appressed: lying flat or pressed against the surface.

Arachnoid: with cobwebby hairs.

Aril: an appendage or outer covering of a seed. Plate 11.

Articulate: a structure with one or more conspicuous breaks in its continuity. Plate 167.

Auriculate: bearing ear-shaped appendages, often at the base of leaves or petals. Plates 169, 203.

Awl: narrowly triangular; gradually tapering upward from a broader base.

Awn: a slender bristle or hair, usually at the tip of a structure. Plate 268b.

Axil: the interior angle between any two structures.

Axillary (lateral) bud: a bud borne in the leaf axil.

Axillary: in an axil.

Basal: arising at the base of the plant or other plant structures.

Berry: pulpy or juicy, multi-seeded, indehiscent fruit. Plates 66, 197.

Bifid: two lobed or cleft, usually at the tip of petals, leaves, or tendrils.

Bipinnate: twice pinnate. Plate 36.

Bisexual: a flower with both stamen(s) and pistil(s).

Bract: a reduced leaf at the base of a flower or an inflorescence. Plates 48, 70, 224.

Bristle: a hairlike projection. Plate 263a.

Bud scales: scales or modified leaves that cover the dormant, growing points.

Bur: a fruiting structure covered with spines or prickles. Plate 140.

Callous: having a hard, thickened texture.

Calyx: the outermost whorl (sepals) of the perianth, usually green or sometimes colored.

Campanulate: bell-shaped.

Capsule: a dry, dehiscent fruit derived from 2 or more carpels. Plates 178, 227, 332.

Carpel: the ovule bearing structure of a flower; a simple pistil or one member of a compound pistil.

Catkin (ament): a bracteate, often unisexual, apetalous, flexible spike-like or cymose inflorescence; the male inflorescence typically falling as a single unit. Plates 37, 56, 312.

Chambered: pith with cross partitions separating empty cavities.

Ciliate: bearing marginal hairs, especially in reference to leaves and bracts.

Cleft: divided into segments to near the middle.

Clone: plants reproducing vegetatively by sharing a common rootstock, as in stolons, runners, rhizomes, etc.

Compound: of 2 or more similar parts.

Compound leaf: a leaf with 2 or more leaflets. Plates 32, 36, 238.

Cone: the seed-bearing structure of gymnosperms, e.g., pine, fir, spruce, etc.

Connate: the fusion or joining together of similar structures. Plate 195.

Cordate: heart-shaped; with a sinus and rounded lobes, often in reference to the base of a structure. Plates 92, 227.

Coriaceous: tough and leathery.

Corolla: collective term for the petals or the whorl(s) of the perianth between the sepals and stamens; usually white or colored.

Corymb: a broad, flat-topped indeterminate inflorescence in which the pedicels are of various lengths and the outermost flowers opening first. Plates 233, 330.

Corymbiform: having the form but not necessarily the structure of a corymb.

Corymbose: resembling a corymb.

Crenate: margins with shallow, round, or obtuse teeth. Plates 51, 284.

Crenulate: diminutive of crenate.

Cuneate: wedge-shaped, usually in reference to leaf bases. Plates 52, 203.

Cuspidate: bearing a strong, sharp point.

Cyme: a determinate inflorescence, often broad and flattened, in which the central flower opens first. Plates 45, 51, 105.

Cymose: resembling a cyme.

Cymule: a small cyme, usually few-flowered.

Deciduous: not persistent or evergreen; falling away.

Decumbent: reclining or lying on the ground, but with the tip turned upright.

Dehiscence: the opening at maturity of anthers or fruits by means of slits, lids, pores, or teeth.

Dentate: with coarse, sharp teeth which project outward. Plate 360.

Denticulate: diminutive of dentate.

Determinate: an inflorescence of limited, definite growth; the terminal flower maturing first and thereby arresting elongation.

Dichotomous: forked into two equal branches.

Dioecious: species having staminate and pistillate flowers on separate plants.

Drupe: a fleshy, indehiscent fruit usually with one seed enclosed in a hard endocarp. Plates 101, 189.

Eciliate: not ciliate.

Eglandular: without glands.

Ellipsoid: a solid that is elliptic in outline. Plate 180 (fruit).

Elliptic: oval; broadest near the middle and gradually tapering to both ends. Plates 182, 349.

Endocarp: innermost layer of the ripened ovary wall.

Entire: a leaf margin without teeth, hairs, spines, etc. Plates 52, 111.

Epidermis: the outermost cell layer.

Even-pinnate: a compound leaf with an even number of leaflets, with the terminal leaflet absent. Plate 155.

Evergreen: bearing thick, leathery leaves throughout the year.

Exserted: extending outward and beyond, as in stamens from the corolla throat or tube.

Exfoliate: to peel off in layers or shreds.

Falcate: sickle-shaped. Plate 262b.

Fascicle: a bundle or dense, close cluster.

Foliaceous: resembling a leaf in texture and appearance.

-foliate: with leaflets.

Follicle: a dry, unicarpellate (having one carpel) fruit which splits along one side at maturity.

Fruit: the mature ovary; accessory structures may be adnate to it.

Gibbous: swollen on one side, usually at the base. Plate 194.

Glabrate: becoming glabrous or smooth with age.

Glabrous: not hairy.

Glandular: having secretory glands or trichomes.

Glaucous: covered with a whitish, waxy substance.

Globose: round, globular, or spherical.

Head: a dense cluster of sessile flowers. Plates 91, 236.

Hemiparasite: a green, photosynthetic plant that draws moisture and other substances from its host.

Herbaceous: plants with little or no woody tissue and dying back to ground level at the end of the growing season.

Hip: the fleshy ripened hypanthium (floral cup) and enclosed achenes (fruits) of *Rosa*. Plate 295.

Hirsute: with coarse or stiff hairs.

Hirsutulous: slightly hirsute.

Hirtellous: minutely hirsute.

Hispid: with long, bristly hairs. Plate 293.

Hyaline: transparent.

Hypanthium: a nearly flat or cup-shaped structure produced from the fusion of sepals, petals, and stamens; it may be free from or adnate to the ovary.

Imbricate: overlapping.

Included: contained within a structure; not exserted.

Indehiscent: not opening at maturity.

Inflated: bladdery, blown up.

Inflorescence: the flower arrangement or mode of flower bearing.

Internode: the portion of the stem between two nodes or points of leaf insertion.

Involucre: one or more series of bracts immediately surrounding a flower cluster(s).

Insipid: without flavor, tasteless.

Lanceolate: lance-shaped; narrow; broadest near the base and tapering to the tip. Plates 108, 153.

Lateral veins: secondary veins arising from the midrib.

Leaf scar: scar tissue remaining on the stem after a leaf has fallen.

Leaflet: an individual unit or secondary "leaf" of a compound leaf.

Legume: a dry fruit from a simple pistil that dehisces along two lines or sutures (e.g., bean family). Plates 92, 155.

Lenticel: openings in the bark of twigs that provide a passage for gas exchange.

Lepidote: covered with small scales. Plate 281.

Linear: long and narrow with parallel sides throughout most of the length. Plates 1, 103.

Lobe: a projecting segment of the leaf margin or other flattened surface.

Margin: the edge of a flat structure, usually a leaf.

Membranous: with a thin, papery texture.

-merous: a suffix denoting the number of parts which constitute a structure.

Midrib: the main or central vein of a leaf; a continuation of the petiole.

Monoecious: with staminate and pistillate flowers on the same plant.

Mucro: an abrupt point or short, spiny tip.

Multiple: a fruit derived from the fusion of the ovaries of numerous flowers; a ripened inflorescence.

Naked bud: a bud surrounded by leaf-like scales that are neither firm nor scaly.

Net venation: a venation pattern where primary and secondary veins form a complex network or reticulum.

Node: the area of a stem where one or more leaves are borne.

Nut: a hard, dry, indehiscent, 1-seeded fruit derived from a 2-many carpellate ovary.

Nutlet: a small nut.

Ob-: a prefix signifying inversion or the reverse of; as in oblanceolate, obovate, obovoid, etc.

Oblong: longer than broad with more or less parallel sides.

Oblanceolate: inversely lanceolate, i.e., narrow at the base and wider at the apex. Plates 171, 280.

Obovate: inversely ovate, i.e., narrow at the base and wider at the apex. Plates 52, 348.

Obsolete: rudimentary or nearly absent.

Obtrullate: widest above the middle and with straight sides; inversely trowel-shaped. Plate 266b.

Obtuse: blunt or rounded at the tip. Plate 86.

Odd-pinnate: a compound leaf with an odd number of leaflets, the terminal leaflet present. Plates 71, 180, 285.

Once-pinnate: a compound leaf with leaflets attached to a central rachis. Plates 71, 285.

Opposite: two leaves at a node, attached directly opposite or essentially so from each other. Plates 24, 130.

Orifice: an opening, mouth, or outlet.

Ovary: the basal, ovule bearing part of the pistil.

Ovate: broadly rounded at the base and narrowed above; shaped like a hen's egg in longitudinal section. Plates 214, 243.

Ovoid: a solid that is oval in outline.

Palmate: lobed or divided in a palm- or hand-like manner. Plates 32, 226.

Panicle: an indeterminate branching raceme; the branches of the primary axis are racemose and the flowers are pedicellate. Plate 33.

Paniculate: resembling a panicle.

Parallel venation: a venation pattern with veins extending in the same direction and equidistant.

Pedicel: the stalk of a single flower of an inflorescence.

Pedicellate: with a pedicel.

Peduncle: the stalk supporting an inflorescence or the flower stalk of species producing solitary flowers.

Peltate: attached away from the margin.

Pendent: drooping or hanging downward.

Pendulous: drooping or hanging downward.

Perennial: plants living for more than two years.

Perfoliate: leaf bases which completely surround the stem, the latter appearing to pass through the former.

Petal: one segment of the corolla.

Petiolate: with a petiole.

Petiole: the stalk supporting the expanded leaf blade.

Petiolule: the stalk of a leaflet of a compound leaf.

Pilose: bearing soft, mostly erect, shaggy hairs.

Pinnate: with leaflets on both sides of a common axis. Plates 145, 285.

Pinnatifid: deeply cut or divided in a pinnate fashion.

Pistil: the ovule bearing portion of a flower, composed of stigma, style (if present), and ovary.

Pistillate: a female flower; bearing a pistil(s), but no functional stamens.

Pith: the soft tissue in the center of stems.

Plumose: resembling a feather; a central axis bearing fine hairs or side branches. Plate 98.

Polygamous: having both unisexual and bisexual flowers on the same plant.

Pome: a fleshy, indehiscent fruit derived from an inferior ovary and surrounded by an adnate hypanthium (e.g., pear, apple, and other related members of the Rosaceae).

Prickle: a small, weak, non-persistent spine-like epidermal projection or the spine-like tips at the end of cone scales in some species of *Pinus*. Plates 16, 18, 47.

Procumbent: lying on the ground, but not rooting.

Prostrate: see procumbent.

Puberulent: minutely hairy or pubescent.

Pubescence: a general term for hairs or trichomes.

Pubescent: covered with soft hairs.

Punctate: with translucent or colored dots, depressions, or pits. Plate 169.

Pyriform: pear-shaped.

Quadrangular: 4-angled or -sided.

Raceme: an elongate, unbranched, indeterminate inflorescence with pedicellate flowers. Plates 44, 99, 178.

Racemose: resembling a raceme.

Rachis: a flower- or leaflet-bearing axis.

Receptacle: the expanded apex of a pedicel upon which the flower parts are borne.

Reflexed: turned downward.

Reniform: kidney-shaped.

Reticulate: resembling a net.

Roseate: rose-colored.

Ruderal: referring to disturbed habitats, e.g., roadsides, pastures, lawns, and other waste areas.

Samara: an indehiscent, single-seeded, dry fruit with a prominent wing (e.g., ash, elm). Plates 147, 257, 347.

Scabrate: slightly rough to the touch.

Scabrous: rough to the touch.

Scale: small modified leaves or bracts.

Scalloped: with a grooved or wavy outline.

Scurfy: covered with scale-like outgrowths.

Semi-: half or hemi-.

Sepal: one of the segments of the calyx.

Sericeous: silky.

Serrate: a margin with sharp, forward pointing teeth. Plates 89, 162.

Serrulate: with small, sharp, forward pointing teeth; diminutive of serrate. Plate 312.

Sessile: attached directly; without a petiole or pedicel.

Sheath: a tubular structure surrounding a stem or other organ.

Shrub: a low, woody plant with one to many slender trunks.

Simple leaf: a leaf not divided into distinct leaflets, the margin may be entire or variously divided. Plates 177, 349.

Silky: with dense, long, straight, appressed hairs giving a silk-like appearance.

Sinus: the space between two lobes or teeth.

Solitary: borne singly.

Spatulate: spoon-shaped. Plate 123.

Spicule: a small, hard projection.

Spine: a sharp, deep-seated outgrowth, not easily removed from the bark; an appendage homologous with a leaf or stipule. Plates 54, 289.

Spike: an elongate, unbranched, indeterminate inflorescence with sessile flowers.

Spur branch: a short, sometimes sharp-pointed branch without internode elongation. Plate 208.

Stalked: buds with a short stem or stalk; not attached directly.

Stamen: the pollen bearing organ of the flower.

Staminate: a male flower; bearing stamen(s), but no functional pistil(s).

Stellate: starlike; trichomes with radiating branches. Plate 233 (inset).

Stipitate: borne on a short stalk.

Stipule: an appendage, usually paired, at the base of a leaf petiole, Plates 61, 314; or adnate to the leaf petiole. Plate 260 (inset).

Stipule scar: a scar remaining on the stem after the stipule has fallen.

Sub-:   a prefix denoting below, slightly, nearly, almost, etc.

Subsessile: nearly sessile, with a very short petiole or stalk.

Subterminal: slightly below the tip.

Subshrub: a weakly woody shrub, or a plant that is primarily woody at the base only. Plates 93, 244.

Succulent: plants having thick, juicy leaves and/or stems. Plate 223.

Suffrutescent: somewhat shrubby; plants with woody bases and herbaceous stems.

Tendril: a modified leaf or stem by which a plant climbs or supports itself. Plates 17, 45, 60.

Terminal: at the apex or tip.

Ternate: in threes, as a leaf divided into three leaflets.

Thorn: a sharp-pointed structure with vascular (woody) tissue; more deeply-seated than a prickle, and more difficult to remove; an appendage homologous with a branch. Plates 155, 238.

Tomentose: with dense, woolly hairs.

Toothed: bearing tooth-like projections, as in the leaf margin.

Translucent: allowing light to pass through, but not transparent.

Tree: a woody plant of considerable stature with one or a few massive trunks and a broad crown.

Trichome: a bristle or hairlike epidermal projection.

Trifid: three-lobed or cleft.

Trifoliate: a plant with three leaves.

Trifoliolate: a compound leaf with three leaflets. Plates 184, 257.

Tripinnate: pinnately compound three times.

Twice-pinnate: pinnately compound two times. Plate 36.

Twig: the stem of the current season.

Umbel: a flat-topped or somewhat rounded indeterminate inflorescence with peduncles or pedicels originating from a common point. Plates 159, 218.

Umbellate: with umbels.

Undulate: wavy, with reference to leaf or perianth margins. Plate 158.

Unisexual: a flower or plant which bears either stamen(s) or pistil(s), but not both.

Urceolate: urn-shaped; wide at the bottom and constricted at the tip.

Vascular bundle scars: the scars remaining on the leaf scar where the broken ends of vascular tissue passed from the stem into the leaf.

Valvate: meeting at the edges but not overlapping.

Variegated: bearing different colors in spots, streaks, etc.

Vine: a plant with weak stems that climb, crawl, or scramble on and supported by other plants or objects.

Viscid: sticky.

Whorl: three or more structures, usually leaves, at a node. Plates 83, 181.

Wing: a dry, flat outgrowth of an organ. Plates 136, 147, 157, 238, 257, 285.

Woolly: with long, soft, intertwined hairs.

Zygomorphic: a flower which may be divided into two equal halves by only one median longitudinal division; sometimes referred to as bilaterally symmetrical.

KEY TO GROUPS

1 Leaves needle-like or scale-like, usually evergreen but deciduous in *Taxodium* (Bald Cypress); plants lacking flowers; seeds borne in cones that are woody or papery (*Pinus*, pines, and others), berry-like (*Juniperus*, Red Cedar), globose with peltate scales (*Taxodium*, Bald Cypress), or single-seeded and surrounded by a fleshy, reddish aril (*Taxus*, Canada Yew) .... Gymnosperms, Page 38

1 Leaves (or leaflets in compound leaves) not needle-like or scale-like, broad and often with various teeth, lobes, or both, usually deciduous (evergreen in a few cases); plants producing flowers; seeds borne in a fruit that may be variously dry or fleshy .... Angiosperms, Page 41

2 Leaves with parallel venation, or if veins not parallel, then a vine with prickles; flowers mostly 3-merous; one species of *Arundinaria* (Cane/Bamboo), two of *Yucca* (Beargrass), and four of *Smilax* (Cat/Saw Brier) are the only woody monocots in our area .... Monocots, Page 41

2 Leaves (or leaflets in compound leaves) with pinnate or palmate veins, i.e., veins never parallel from leaf base to apex; flowers commonly 4- or 5-merous or with numerous parts; more than 95 percent of our woody flora belongs here .... Dicots, Page 42

GYMNOSPERMS

1 Leaves needle-like, in clusters of 2–5 and enclosed in a basal (sometimes early deciduous) sheath . . . . *Pinus*
1 Leaves linear, scale-like, or awl-shaped, and lacking a basal sheath 2
   2 Leaves opposite, scale-like, or awl-shaped on young branches, mostly <3.5 mm long 3
      3 Cone woody, seeds winged; twigs flattened . . . . *Thuja*
      3 Cone berry-like, seeds wingless; twigs 4-angled . . . . *Juniperus*
   2 Leaves alternate, linear, and >3.5 mm long 4
      4 Leaves (and branches deciduous); unopened cones globose, the scales peltate; plants of wetlands . . . . *Taxodium*
      4 Leaves evergreen; unopened cones longer than wide, the scales flattened and not peltate, or scales absent and the seed surrounded by a fleshy, red aril; plants of drier sites 5
         5 Leaves 4-angled . . . . *Picea*
         5 Leaves flat 6
            6 Leaves abruptly acuminate at the tip; seed single and enclosed in a fleshy, reddish aril . . . . *Taxus*
            6 Leaves rounded at the tip; seeds borne in woody or papery cones 7
               7 Cones pendent, <4 cm long; leaves with a short, petiole-like base . . . . *Tsuga*
               7 Cones erect, >4 cm long; leaves sessile . . . . *Abies*

Pinaceae   *Abies* (Fir)

A. fraseri (Pursh) Poir., Fraser F./She-Balsam; medium trees; monoecious; both male (early summer) and female (fall) cones produced on the underside of branches of the previous season; high elevation forests >4500 ft; infrequent; U. **Threatened.** Commercially important for plantings and Christmas trees. The balsam woolly adelgid, introduced from Europe in 1908, began attacking this species around 1960. This insect, combined with acid precipitation, has had a devastating impact on the high elevation spruce-fir forests; almost all mature fir trees are dead or dying. Seedlings are plentiful but the long-term recovery potential is uncertain. Plate 1.

Cupressaceae   *Juniperus* (Cedar)

J. virginiana L., Red C.; shrubs or small trees; mostly dioecious; cones terminal or axillary, male yellowish (early spring), female bluish black and berry-like (late fall); dry, poor soils, especially over limestone, successional fields, bluffs; common;

statewide. Important as the "Christmas tree" of rural mid-South, it is also used for fenceposts, as ornamentals, and for furniture, especially cedar chests. Plate 2.

Pinaceae   *Picea* (Spruce)

P. rubens Sarg., Red S./He-Balsam; large trees; monoecious; cones pendent, both male (early summer) and female (fall) appearing on axillary buds of the previous season; high elevation forests; infrequent; U. Red Spruce is shallow-rooted, and the decline of Fraser Fir, combined with acid precipitation, has made it more susceptible to wind throw. Plate 3.

Pinaceae   *Pinus* (Pine)

Medium to large trees; monoecious; male cones axillary (spring), in clusters at base of current season's growth, female cones subterminal or axillary (maturing and releasing seeds in the second season but often remaining persistent for several years).

1 Leaves 5/bundle; basal sheath early deciduous; cone scales without a
   prickle at the apex (soft pine) . . . . *P. strobus*
1 Leaves 2–4/bundle; basal sheath persistent; cone scales with a prickle at
   the apex (hard pines) 2
   2 Leaves >13 cm long . . . . *P. taeda*
   2 Leaves <13 cm long 3
      3 Leaves 2/bundle, rarely 3/bundle 4
         4 Cones asymmetric, scales with stout, curved
            prickles . . . . *P. pungens*
         4 Cones symmetric, scales with short, straight prickles 5
            5 Leaves strongly twisted, 3–8 cm long . . . . *P. virginiana*
            5 Leaves straight or slightly twisted, 7–13 cm
               long . . . . *P. echinata*
      3 Leaves 3/bundle, rarely 2 or 4/bundle 6
         6 Leaves 2 or 3/bundle, flexible, straight or slightly twisted;
            cone short-stalked . . . . *P. echinata*
         6 Leaves 3 or 4/bundle, rigid, straight; cone
            subsessile . . . . *P. rigida*

P. echinata Mill., Shortleaf P.; old fields and upland woods; frequent; M and E TN. Important for lumber and pulpwood; seeds are eaten by wildlife. The common name may be confusing since the needles are longer than those of Virginia Pine. Plate 4.

P. pungens Lamb., Table Mountain P.; dry, rocky slopes and ridges; infrequent; E TN. Of little commercial value. Plate 5.

P. rigida Mill., Pitch P.; dry upland woods; infrequent; VR, U. Occasionally used for lumber and pulpwood. Plate 6.

P. strobus L., White P.; moist or dry slopes and ridges; frequent; E TN and disjunct on the WHR. Commercially important for lumber, Christmas trees, and landscaping. Plate 7.

P. taeda L., Loblolly P.; old fields, roadsides, and dry woods; frequent; originally native to the statewide southern tier of counties except MV, now planted and spreading throughout. Commercially important for lumber and pulpwood. Plate 8.

P. virginiana Mill., Virginia/Scrub P.; fields and dry woods; common; statewide, but rare on the EHR and in W TN. Occasionally used for pulpwood. Plate 9.

Taxodiaceae   *Taxodium* (Bald Cypress)

T. distichum (L.) Rich.; medium to large trees; monoecious; male cones (spring) in pendent panicles and conspicuous during the winter prior to pollination, female cones nearly round (fall); swamps, bottomlands, riverbanks; common in W TN and occasionally along the eastern shorelines of the Tennessee River (WHR), planted elsewhere across the state in damp soils. Highly prized for lumber and interior trim; also resistant to decay and excellent for decking and where structures are in contact with soil. Shallow roots, where the water level fluctuates, often produce upright, conical structures called "knees." The function of these outgrowths, if any, is subject to debate. Plate 10.

Taxaceae   *Taxus* (Canada Yew)

T. canadensis Marshall; small shrubs; dioecious; both male (spring) and female (fall) cones axillary; the female cone is surrounded by a fleshy, reddish aril; dry, shady woods at base of bluff lines; Pickett Co. (CU). **Endangered.** Plate 11.

Cupressaceae   *Thuja* (Arbor Vitae/White Cedar)

T. occidentalis L.; shrubs or small trees; monoecious; both male (spring) and female (fall) cones terminal; damp, calcareous seeps and slopes; infrequent; northern counties of the EHR and E TN. **Special Concern.** Resistant to insects and fungi and used for lumber, fence posts, and shingles. Plate 12.

Pinaceae   *Tsuga* (Hemlock)

Medium to large trees; monoecious; male cones (spring) axillary, female cones (fall) terminal. Widely used as ornamentals, lumber, and pulpwood; the bark was once used for tanning. The hemlock woolly adelgid has moved recently into the southern Appalachians and has the potential of destroying many stands.

1   Leaves <1.5 cm long, mostly arranged in one plane, the margins minutely serrulate; seed cones <2 cm long . . . . *T. canadensis*
1   Leaves mostly >1.5 cm long, spirally arranged, the margins entire; seed cones >2 cm long . . . . *T. caroliniana*

T. canadensis (L.) Carrière, Eastern/Canada H.; moist woods and streambanks; frequent; EHR and E TN. Used as an ornamental, especially for hedges; the wood is of poor quality for lumber. Plate 13.

T. caroliniana Engelm., Carolina H.; rocky woods and bluffs; infrequent; upper U. **Threatened.** Plate 14.

ANGIOSPERMS: MONOCOTS

1   Vines climbing by tendrils; stems usually prickly . . . . *Smilax*
1   Plants erect, without tendrils; stems not prickly 2
    2   Stems slender and cane-like, usually several feet tall and with prominent nodes; internodes hollow; leaf bases forming sheaths around the stem; leaf blades relatively membranous . . . . . . *Arundinaria*
    2   Stems short, rarely >1 ft tall and with non-prominent nodes; internodes not hollow; leaves spirally arranged and basally clustered; leaf blades sword-like and with sharp-pointed tips . . . . *Yucca*

Poaceae   *Arundinaria* (Cane/Switch Cane/Bamboo)

A. gigantea (Walter) Muhl.; mostly evergreen, clonal shrubs; flowers and fruits in terminal and lateral panicles (spring–summer) but rarely flowering; bottomland woods and thickets, streambanks, ravines; common; statewide. Other genera of exotic bamboos sometimes appear in cultivation. *A. gigantea* ssp. *tecta* (Walter) McClure, *A. gigantea* ssp. *macrosperma* (Michx.) McClure. Plate 15.

Smilacaceae   *Smilax* (Cat/Saw Brier)

Vines climbing by tendrils; dioecious; flowers greenish, in umbels (spring–early summer); fruit a bluish black berry (summer–fall). Our only woody vines with prickles, which often are most abundant at the base of the plant.

1 Leaves glaucous beneath .... *S. glauca*
1 Leaves green beneath 2
  2 Leaf margins entire; stems distinctly 4-angled, especially on young
    growth .... *S. rotundifolia*
  2 Leaf margins rarely entire, usually ciliate-serrulate or with prickles;
    stems round or slightly 4-angled 3
    3 Prickles slender, needle-like; stems round or only slightly angled;
      leaves not lobed, thin, the margin ciliate-serrulate ..... *S. tamnoides*
    3 Prickles stout, flattened; stems slightly 4-angled; leaves thick, often
      lobed at the base, the margin usually with prickles .... *S. bona-nox*

S. bona-nox L.; dry woods, fields, cutover sites; common; statewide. Plate 16.

S. glauca Walter; woodlands and ruderal sites; common; statewide. Plate 17.

S. rotundifolia L.; woodlands, thickets, and open sites; common; statewide. Plate 18.

S. tamnoides L.; woodlands and thickets, less common on disturbed sites than the previous species; common; statewide. *S. hispida* Muhl. Plate 19.

Agavaceae   *Yucca* (Yucca, Spanish Bayonet, Beargrass)

Low shrubs; flowers white, in large terminal panicles (early summer), fruit a capsule (fall). Pollinated only by the Pronuba Moth; the larvae hatch inside the ovary and feed on the seeds. Often used in ornamental plantings.

1 Branches of the inflorescence densely pubescent .... *Y. flaccida*
1 Branches of the inflorescence glabrous .... *Y. filamentosa*

Y. filamentosa L.; dry woods, roadsides, and along streambanks; infrequent; M and E TN. Plate 20.

Y. flaccida Haw.; same habitats as the preceding; infrequent; statewide. This taxon is closely related to and has been recognized as a variety and as a synonym of *Y. filamentosa*. *Y. filamentosa* L. var. *smalliana* (Fernald) H.E.Ahles.

ANGIOSPERMS: DICOTS

1 Vines, climbing or twining on other plants .... Key A, Page 43
1 Plants not vines 2
  2 Plants erect, trailing, decumbent, prostrate, or mat forming; often
    rooting at the nodes; rarely >2 dm tall .... Key B, Page 45

Key A

Vines (climbing or twining on other plants)

1 Leaves simple 2
  2 Leaves opposite 3
    3 Stems climbing by aerial roots . . . . *Decumaria*
    3 Stems trailing or climbing by twining 4
      4 Stipules or their scars absent . . . . *Lonicera*
      4 Stipules or their scars present, sometimes minute (examine
         carefully at 10x) 5
         5 Sap milky (examine broken petiole base); leaves
            short acuminate, pubescent beneath; corolla <1 cm
            long . . . . *Trachelospermum*
         5 Sap not milky; leaves acute, glabrous beneath; corolla
            >1 cm long . . . . *Gelsemium*
  2 Leaves alternate 6
    6 Lower and sometimes upper portion of stem with
      prickles . . . . *Smilax*
    6 Plants without prickles 7
      7 Plants climbing by tendrils 8
        8 Leaves entire and not lobed; tendrils at the tips of lateral
          branches . . . . *Brunnichia*
        8 Leaves toothed or lobed or both; tendrils opposite
          the leaves 9
          9 Stems of the previous season with white pith;
             inflorescence cymose . . . . *Ampelopsis*
          9 Stems of the previous season with brown pith;
             inflorescence paniculate . . . . *Vitis*
      7 Plants without tendrils 10
        10 Plants climbing by aerial roots . . . . *Hedera*
        10 Plants without aerial roots 11
          11 Leaves palmately veined 12

24 Plants climbing by aerial roots (**caution: poison
ivy**) . . . . *Toxicodendron*
24 Plants without aerial roots . . . . *Pueraria*
23 Leaflets more than 3 25
25 Plants with tendrils; leaves bi- or tri-pinnately
compound . . . . *Ampelopsis*
25 Plants without tendrils; leaves once-pinnately
compound . . . . *Wisteria*

Key B

Plants Erect, Trailing, Decumbent, Prostrate, or Mat Forming; Often Rooting at the Nodes; Rarely >2 dm Tall

1 Stems flat, fleshy; leaves minute or absent (cactus) . . . . *Opuntia*
1 Stems not flat and fleshy; leaves present 2
2 Leaves alternate, or rarely appearing whorled 3
3 Leaves simple 4
4 Stems prostrate or plants with both prostrate and
upright stems 5
5 Stems densely hispid; leaves >1 cm wide . . . . *Epigaea*
5 Stems glabrous or essentially so; leaves <1 cm
wide . . . . *Vaccinium*
4 Stems erect 6
6 Leaf blades <2.5 cm long . . . . *Buxella*
6 Leaf blades (the larger ones) >2.5 cm long 7
7 Leaves with the odor of wintergreen, not
variegated . . . . *Gaultheria*
7 Leaves without the odor of wintergreen, variegated
white or gray along the midrib and some lateral
veins . . . . *Chimaphila*
3 Leaves compound 8
8 Stems with prickles . . . . *Rubus*
8 Stems without prickles . . . . *Potentilla*
2 Leaves opposite or sub-opposite 9
9 Leaves <1 cm wide 10
10 Leaves linear 11
11 Leaves pubescent beneath, <2.5 mm wide, minty
aromatic, glandular (10x) . . . . *Conradina*

    11 Leaves glabrous beneath, >2.5 mm wide, not aromatic
        or glandular (10x) . . . . *Paxistima*

  10 Leaves ovate, oblong, or oblanceolate 12

    12 Leaves leathery, ovate to oblong, mostly <1 cm long;
        petals 5 . . . . *Leiophyllum*

    12 Leaves not leathery, oblanceolate, mostly >1 cm long;
        petals 4 . . . . *Hypericum*

 9 Leaves >1 cm wide 13

  13 Stems mostly prostrate, trailing; upright branches, if present,
    <1 dm tall 14

    14 Leaves, at least some, toothed above the middle;
        fruit dry . . . . *Linnaea*

    14 Leaves strictly entire; fruit a fleshy, red berry . . . . . . *Mitchella*

  13 Stems both trailing and with upright branches
    >1 dm tall 15

    15 Leaves entire; sap milky (examine broken
        petiole base) . . . . *Vinca*

    15 Leaves toothed; sap not milky . . . . *Euonymus*

Key C
LEAVES SIMPLE AND ALTERNATE

1 Leaf margin entire, occasionally ciliate but not regularly producing
teeth or lobes or undulations . . . . Key 1

1 Leaf margin variously toothed (at least in part), lobed, or
undulate . . . . Key 2

Key 1

1 Leaves evergreen (thick and leathery) or tardily deciduous, the lateral
veins often obscured 2

 2 Stipules or their scars completely encircling the twig . . . . . . *Magnolia*

 2 Stipules or their scars absent 3

  3 Mature leaf blades <3 cm wide . . . . *Vaccinium*

  3 Mature leaf blades >3 cm wide 4

    4 Leaves with peltate scales beneath (10x) 5

      5 Leaves rusty-brown scaly beneath; thorns absent;
        native . . . . *Rhododendron*

      5 Leaves silvery and usually dotted with a few brown scales
        beneath; thorns often present; introduced . . . . *Elaeagnus*

    4 Leaves not scaly beneath 6